CLIMATE ADAPTATION INVESTMENT PLANNING

LESSONS LEARNED

APRIL 2026

ASIAN DEVELOPMENT BANK

ADB

6 ADB Avenue, Mandaluyong City, 1550 Metro Manila, Philippines
Tel +63 2 8632 4444; Fax +63 2 8636 2444
www.adb.org

ISBN 978-92-9277-699-2 (print); 978-92-9277-700-5 (PDF); 978-92-9277-701-2 (ebook)
Publication Stock No. SPR260072-2
DOI: http://dx.doi.org/10.22617/SPR260072-2

Note:
In this publication, "$" refers to United States dollars.

Cover design by Lowil Fred Espada

On the cover: ADB supports small and medium farmers in Nepal to increase their household income and resilience (photo by Narendra Shrestha).

CONTENTS

TABLES, FIGURES, AND BOXES

Tables

Figures

Boxes

ACKNOWLEDGMENTS

This report was prepared by Julie Dekens, principal researcher at the International Institute for Sustainable Development (IISD), together with Asian Development Bank (ADB) consultants Paul Watkiss, John Ward, Robert Wilby, Nand Agrawal, Rathana Peou, David Manukyan, Michael Alan Schur, Nur Kholis, Vardan Melikyan, Batkhuyag Choijiljav, Munkhzul Chimid-Ochir, Siveun Nhak, Choikhand Janchivlamdan, and Sonam Wangdi, Abid Hussain, economies lead at the International Centre for Integrated Mountain Development (ICIMOD), and the following ADB staff: Arghya Sinha Roy, director (Adaptation and Resilience); Nanki Kaur, senior adaptation solutions specialist; Natasha Kuruppu, former climate change specialist; Kyla May Matias, climate change officer; Vardan Karapetyan, principal climate change officer; Arun Shumshere Rana, principal project officer; Putra Dwitama, senior climate change officer; Enkh-Amgalan Tseelei, senior climate change officer; and Thavin So, senior climate change officer.

The report benefited greatly from review comments received from Anne Hammill, associate vice-president, Resilience, and Maribel Hernandez, senior policy advisor, Adaptation Finance from IISD; as well as Christina Cheong, low carbon solutions specialist (Urban) and Heejin Lee, climate change specialist from ADB. It has been enriched by guidance provided by the governments of Armenia, Bhutan, Cambodia, Indonesia, Mongolia and Nepal; ADB Sector Groups (Agriculture, Food, Nature and Rural Development, Finance, Public Sector Management and Governance, and Water and Urban Development); and ADB Resident Missions (Armenia, Bhutan, Cambodia, Indonesia, Mongolia, and Nepal).

The team thanks Sugar Gonzales, senior climate change officer (Climate Change Adaptation); Evangeline Sucgang, technical assistance coordinator; James Unwin, editor; Lowil Espada, graphic designer; Amelita de Dios, associate operations officer; and Marilyn Parra, operations analyst, for their editorial and coordination support. The team also acknowledges the support provided by Anna Liza Cinco, Ailen Marylou Bigay, colleagues at ICIMOD, Chhun Chhim, Thakur Devkota, Josephine Domingo, and Nancy Bustamante in the implementation of the Climate Adaptation Investment Planning (CAIP) program.

This publication has been developed with the support of the CAIP program implemented by the ADB in partnership with the International Institute for Sustainable Development and International Centre for Integrated Mountain Development, with financing from ADB, the Climate Investment Funds, and the Government of Ireland.

ABBREVIATIONS

ADB	Asian Development Bank
CAIP	Climate Adaptation Investment Planning
COP	Conference of the Parties
MDB	multilateral development bank
NAP	national adaptation plan
NDC	nationally determined contribution
UNFCCC	United Nations Framework Convention on Climate Change

A woman working the rice field in Madhesh Province, Nepal. © Nabin Baral, ICIMOD

EXECUTIVE SUMMARY

As climate change intensifies its global impacts on communities, ecosystems, and economies, developing countries must urgently implement their adaptation priorities to tackle both present and future challenges. This report demonstrates how national adaptation priorities can be translated into investment-ready pipelines that more effectively mobilize public and private finance, while unlocking broad development benefits.

The current speed and scale of adaptation financing and implementation remain insufficient. Limited understanding of fiscal and financial risks, fragmented project-based approaches, and inadequate investment planning have hindered systemic, locally responsive adaptation efforts. How developing countries can mobilize public and private investments in adaptation more strategically—at the required speed and scale—to implement national adaptation priorities becomes a critical question.

A wave of internationally supported initiatives and frameworks in recent years shows strategies gaining ground to help countries mobilize public and private investments in adaptation. Most encourage countries to use adaptation priorities from national adaptation plans (NAPs) and nationally determined contributions (NDCs) to attract finance.

This report identifies practical insights on mobilizing investments for national adaptation priorities from the Climate Adaptation Investment Planning (CAIP) program of the Asian Development Bank (ADB). CAIP is designed to support national processes to scale up investment in adaptation and bridge gaps between planning and financing, helping actors to realize opportunities across different scales. ADB and its participating developing member countries co-designed a conceptual framework to operationalize CAIP and identify policy areas needing further investment.

The report is informed by case examples, country reports, and outcomes from regional and national dialogues from 2022 to 2025 in six countries: Armenia, Bhutan, Cambodia, Indonesia, Mongolia, and Nepal. Building on lessons learned, the CAIP program is now expanding to seven other countries: Bangladesh, India, the Lao People's Democratic Republic, Palau, the Philippines, Sri Lanka, and Timor-Leste.

The CAIP framework comprises five steps for developing countries to translate their national adaptation priorities into investment-ready programs.

1. **Context and Rationale:** Review national development and adaptation plans, fiscal frameworks, and public investment systems to align adaptation investments with government priorities and financing opportunities.

2. **Strategic Climate Diagnostic:** Identify key present and future climate risks using existing assessments and proxies to inform strategic investment priorities.

3. **Prioritization and Sequencing:** Group and sequence adaptation actions into investment packages based on urgency and participatory input.

4. **Economic and Financial Rationale:** Assess costs, benefits, and returns to justify investments and identify suitable finance sources.

5. **Financial Matchmaking:** Match investment packages to diverse finance sources and financial instruments, and explore enabling policy reforms.

This framework emphasizes a supportive enabling environment—strong government ownership, institutions, capacity, and improved awareness among domestic finance institutions and private actors. It addresses global challenges in mobilizing climate adaptation investments by aligning financing with national development goals and identifying systems-level, cross-sector adaptation packages. It highlights a path to accelerating speed and scale by prioritizing urgent risks, leveraging diverse finance sources, and engaging ministries of finance to integrate adaptation planning into public financial management systems to support transformative adaptation across various systems. The approach also aims to strengthen programmatic coherence and resources at the local level through tailored packages and decentralized governance.

Early application of the CAIP framework across the six countries shows that most adaptation investment packages yield strong economic returns, though the financial viability of some packages remains limited and will often require public finance. Private investment potential varies by sector and package, with early opportunities in agricultural value chains and longer-term prospects in areas like climate services and water sector partnerships. Diverse financing modalities—including blended instruments and policy-based lending—are essential to operationalize comprehensive adaptation investment plans.

Closer examination of these country experiences allows nine lessons to be distilled.

Lessons learned

Adaptation investment planning must operate within the existing development planning and fiscal context of the country, requiring a careful balance between ambition and pragmatism.

Persistent development barriers hindered the ability to mobilize adaptation investments in a coordinated and effective way in the six countries. Aligning with these realities proved essential to ensure a pragmatic approach that balanced ambition with the constraints of national development planning and budgeting. In Mongolia, for instance, it was crucial for the adaptation investment planning process to be aligned with budget cycles and the planning of the public investment program.

A focused and iterative approach is essential for adaptation investment planning.

Given the complexity and details required, simultaneously preparing every adaptation priority outlined in a national adaptation plan (NAP) document or other strategic documents for investment was found to be both impractical and ineffective. Instead, countries selected a manageable unit of focus—a sector and/or specific region, depending on national priorities and governance structures. Countries adopted an iterative approach that allowed continuous refinement of investment packages. Multiple rounds of prioritization and sequencing—guided by climate risk analysis, financial appraisal, and matchmaking results—helped ensure coherence and broadened financing opportunities.

Tailoring climate risk diagnostics to specific adaptation priorities helps prioritize investments.

All countries had already conducted some form of climate risk assessment to identify their adaptation priorities. Yet, all required more focused and up-to-date analyses of climate risks linked to the adaptation priorities for the selected sector, subsector, or region and considering overlooked aspects such as multi-hazard analysis, cross-sector risk linkages, and uncertainties. The climate risk diagnostics therefore complemented existing information and provided more specific data needed to further prioritize and detail particular adaptation priorities. In Cambodia, results revealed heat risks to labor productivity—an aspect that had been omitted from the agriculture sector NAP document and was subsequently added to an adaptation investment package.

Comprehensive adaptation investment packages address multiple climate risks to development outcomes at a system level and include different types of investments.

Identifying and prioritizing adaptation investment packages often required refinement to specify financing sources. Some NAP documents mention broad priorities, such as "improve water security" and "climate-smart agriculture", while others signpost specific technologies. However, the source and approach to financing require detail as well as consideration of the implementation modality (finance required for extension services, or access to finance for technology). In turn, these need to be packaged up at scale to provide programmatic investments. Countries' experiences highlight the need and feasibility of grouping adaptation priorities into comprehensive and complementary investment packages for specific sectors to address adaptation in an integrated way.

Appraising the benefits and viability of adaptation investment packages can facilitate their integration into public management processes and identify opportunities for the private sector.

Countries are increasingly quantifying adaptation costs through their NAP processes. However, most still lack analysis of the benefits. Results from applying the Climate Adaptation Investment Planning (CAIP) framework showed that it was feasible and valuable to conduct qualitative and semi-quantitative assessments of the benefits to build the case for investing in adaptation—for example, through the relevant public investment management process in government or the development of concept notes for potential support from a multilateral development bank. Light-touch benefits appraisal was complemented with indicative analysis of the potential economic and financial returns. More detailed quantitative appraisals will still need to be completed later during project preparation. For some adaptation investment plans, which focused on a smaller number of packages (such as Armenia and Indonesia), more detailed cost-benefit analyses provided specific economic and financial internal rates of return.

Decision trees serve as a practical starting point for matching adaptation priorities with financing sources and instruments, guiding countries toward actionable steps.

For some countries, granular, context-specific financial matchmaking was essential to align refined adaptation investment packages with a varied mix of financing sources and instruments—ensuring coherence with national priorities and clarifying who would finance, fund, implement, or support each investment. Heuristics such as decision trees can be a useful entry point for undertaking this matchmaking.

Comprehensive adaptation investment packages and targeted finance matchmaking require broadening and deepening engagement with key actors.

Sustained engagement—strengthened through actor mapping across ministries, agencies, and financial institutions—was essential for building consensus, ensuring ownership, and identifying the sources and opportunities for mobilizing diverse financing. In Cambodia, for example, having the Ministry of Economy and Finance as a co-implementer of adaptation investment planning from the outset helped strengthen participation by line ministries.

Country-led approaches yield coherence in adaptation investment planning across different actors, sectors, geographies, and governance levels.

Countries were able to leverage existing institutional mechanisms to facilitate collaboration, or to create new ones. For example, in Bhutan, establishment of an interagency working group brought together ministries and departments responsible for water management—including in agriculture, water and sanitation, energy, and flood management. This fostered strong government ownership of the overall process while ensuring the sustainability of the CAIP work beyond the Technical Assistance program.

Adaptation investment planning requires specialized skills and adequate human and financial resources.

The CAIP process, as a new approach, requires ongoing awareness-raising and capacity and institutional strengthening across line ministries and departments, as well as technical assistance. Preparing adaptation priorities for investment in a specific sector or region typically requires significant resources over an extended time frame. The timeline can vary based on factors like the availability of quality climate assessments, the need for pre-feasibility studies, and the status of adaptation planning and coordination within the government. Interdisciplinary expertise in the fields of strategic climate risk analytics, adaptation economics, climate finance, and investment planning is critical. In addition, forming partnerships with national institutions enhanced the process by facilitating access to relevant actors and data and information.

Strengthening CAIP engagement

Successful implementation of climate adaptation investment plans will necessitate ongoing coordination and support, potentially through project preparation facilities and country climate platforms, emphasizing strong government ownership and collaboration and improved coordination with development partners.

As countries advance through NAP cycles, documenting CAIP's impact will be vital to continue to learn lessons and improve the process, and to determine how effective this approach has been at mobilizing increased volumes of diverse finance sources. The immediate priority involves sustaining progress through feasibility studies that secure financing.

Future efforts must strengthen gender equality and social inclusion in investment planning, ensuring adaptation addresses the needs of marginalized groups. This should be enhanced to ensure that adaptation investment benefits all groups in society, especially when moving to new sources of finance (especially private) to ensure this does not further exacerbate inequalities. Engagement should extend beyond finance ministries to include civil society, as well as philanthropic funding, particularly for low-return investments.

Countries also need support mobilizing domestic resources through fiscal tools while attracting private investment through blended finance and policy reform.

1 INTRODUCTION

The impacts of climate change on communities, ecosystems, and economies worldwide are expected to intensify. Developing countries urgently need to implement their climate adaptation priorities to deal with both current and future impacts.

In 2025, three milestones underscored the challenges associated with the climate crisis. First, in June, at the Fourth International Conference on Financing for Development, countries recognized that societies have reached a critical crossroads—a period marked by profound transformation, geopolitical tensions, conflicts, and rising economic challenges (United Nations 2025). Then, in September, the second scientific assessment on planetary boundaries concluded that seven out of the nine planetary boundaries have now been breached, thereby undermining the systems that support life on Earth (PBScience 2025). Finally, through the United Nations Framework Convention on Climate Change (UNFCCC) in November, the Conference of the Parties (COP) presidencies of Azerbaijan (COP29) and Brazil (COP30) released the Baku to Belem Roadmap to mobilize climate finance for developing countries by 2035. The road map identifies that scaling up investments in support of adaptation require governments to accelerate the move from "fragmented projects to programmatic pipelines integrated into fiscal frameworks" (UNFCCC 2025a).

In this context, it is critical for developing countries to adopt novel approaches to financing adaptation by leveraging diverse resources beyond international public finance. Such approaches should aim to align with and strengthen domestic priorities, resources, and processes. Both public and private investments in adaptation are urgently needed to curtail greater economic and noneconomic losses from climate-related impacts in future. Understanding adaptation not only as a cost of reducing risks but also as an investment that delivers benefits will be vital (UNFCCC 2025b). These benefits are wide-ranging, from improved well-being and job creation to enhanced fiscal health. They can enable countries to achieve their national development goals and priorities.

This report explores how developing countries can move from planning to implementing their national adaptation priorities through faster, greater, and smarter mobilization of public and private investments. The report contributes to global discourse on effective scaling up of investments in adaptation. It is designed to offer practical insights and key learnings from the Climate Adaptation Investment Planning (CAIP) program, a regional initiative supported by the Asian Development Bank (ADB)—and identified as an example in the *Report on the Baku to Belem Roadmap to 1.3T*—to help countries scale up implementation of adaptation in a programmatic manner (UNFCCC 2025a).

The report aims to inform governments across developing countries, development practitioners, as well as the private sector and other key actors involved in programming and implementing adaptation activities at different scales, from local to national.

Under the CAIP program, ADB and its participating developing member countries and experts co-designed a conceptual framework to operationalize the CAIP program in six countries in Asia and the Pacific between 2022 and 2025: Armenia, Bhutan, Cambodia, Indonesia, Mongolia, and Nepal. CAIP is led by the countries and initial support provided through targeted technical assistance from ADB and the Climate Investment Funds. Drawing on practical case examples, country reports, and outcomes of regional and national dialogues, this report presents key cross-country learnings and results while highlighting the policy areas that need further investments. Building on emerging lessons from the six countries, the CAIP program, with support from ADB and the Government of Ireland, is now applying the same approach to seven other countries: Bangladesh, India, the Lao People's Democratic Republic, Palau, the Philippines, Sri Lanka, and Timor-Leste.

Key terms used in this report are defined in Box 1. This report is structured as follows. Section 2 discusses the six core global challenges in mobilizing climate adaptation investment. Section 3 provides an overview of global and regional initiatives aimed at responding to these challenges, introduces the CAIP conceptual framework developed by countries and ADB, and presents the key results from its application in the six countries. Section 4 distils emerging lessons from their experiences and concludes by outlining potential next steps for advancing progress.

Box 1: Definition of Key Terms in This Report

Adaptation investment: Public and private financial resources allocated to climate adaptation activities with the expectation of generating benefits. Benefits may include economic and noneconomic gains (such as job creation and enhanced well-being), fiscal health benefits (such as reduced expenditure and increased revenue from climate-resilient economic activities), and financial returns (namely, revenue generation).

Adaptation investment package: A set of adaptation activities related to a sector, region, or theme, or to financing sources. Adaptation investment packages aim to cluster or bundle adaptation activities to build resilience at scale and at a system level.

Adaptation investment readiness: A country or entity demonstrates investment readiness when it can develop investment packages, articulate their value proposition, and identify the appropriate financing models to enable investment in adaptation activities.

Climate adaptation investment planning (CAIP): The CAIP framework supports decision-makers to strengthen (a) system-level risk analytics to develop programmatic adaptation investment packages; (b) economic and financial analysis to articulate the return on investment in adaptation and allocate investment responsibility across public and private sectors; (c) finance matchmaking to leverage public and private finance for investment in adaptation; and (d) the policy and regulatory environment that enables investment in adaptation. The CAIP framework is guided by public financial management systems to embed adaptation investments in economic planning and budgeting.

National adaptation plan (NAP): A domestic process established under the United Nations Framework Convention on Climate Change in 2010 that aims to systematically integrate medium- and long-term adaptation considerations into development decision-making. The ultimate goal is to achieve climate-resilient development. A key milestone in the NAP process is the identification of national adaptation priorities, which are articulated in a NAP document or other strategic instruments such as nationally determined contributions (NDCs). For clarity, this report refers to either the process as the "NAP process" or the document identifying adaptation priorities as the "NAP document". For more information, see Box 2.

2 GLOBAL CHALLENGES IN MOBILIZING CLIMATE ADAPTATION INVESTMENT

All countries face six core challenges in mobilizing investment in adaptation at the required speed and scale needed to address climate risks:

1. **Climate risks will continue to intensify and threaten development progress.**

The adverse impacts of climate change already undermine socioeconomic progress, particularly in poverty reduction, across many countries. Observable effects include disruptions to food production, water security, human health and well-being, infrastructure, and settlements, which collectively hinder efforts to achieve the Sustainable Development Goals (IPCC 2023). Even under low-emission scenarios, projected climate risks are expected to be more complex and severe than those currently experienced, meaning that countries need to scale up investment in adaptation now to avoid higher future losses.

2. **The speed and scale of adaptation financing and implementation remain insufficient.**

Speed refers to the pace at which developing countries are mobilizing public and private investment for adaptation, which lags far behind the urgent need to implement adaptation priorities (UNEP 2023). Scale refers to the insufficient volume of finance available for adaptation relative to countries' needs. Investment in adaptation is estimated to cost between $310 billion and $365 billion annually in developing countries by 2035, in 2023 prices (UNEP 2025). This far exceeds the total tracked adaptation finance of $65 billion in 2023 (CPI 2025), highlighting the magnitude of the challenge. Adaptation project proposals have typically focused narrowly on a single source of finance—often international public funds—and a small number of financial instruments, grants, or concessional loans, rather than on mobilizing diverse financing sources and financial instruments.

3. **Understanding of the fiscal and financial risks of climate change and adaptation is limited.**

To date, most ministries of finance and development planning bodies, along with central banks, lack a full understanding of climate-related fiscal and financial risks, including impacts on governments' revenues and expenditures, and thus the impact on public finances. They also have not factored in the fiscal impacts of likely increased adaptation spending, much of which will fall to the public budget (ADB 2024a). As of February 2025, only 9 NAP documents out of 59 refer to climate-resilient fiscal planning (Hernández, Ledwell, and Yang 2025). Some good practices are emerging, but in general adaptation planning and financing has largely been a stand-alone exercise, often undertaken outside of finance ministries, and with limited integration of adaptation priorities in public financial management systems, including annual and medium-term development planning, budgeting, and procurement. This has resulted in fragmented, misaligned, and incremental adaptation responses, and a failure to align wider financial flows toward climate-resilient development.

4. **Programmatic approaches to adaptation are lacking.**

To date, climate adaptation measures have primarily been implemented as separate, ad hoc initiatives with limited potential for transformational change. This project-based approach has primarily focused on a single sector and often addressed only one climate hazard, and/or targeting small geographic areas. It typically involves a limited set of actors and concentrates on short-term adaptation needs. As a result, adaptation projects often fail to tackle the root causes of climate vulnerability, which can reinforce social inequalities and marginalization while weakening governance systems over time. This fragmented approach hampers financing adaptation at scale and is inadequate for addressing the scale and speed of climate risks, as long-term resilience requires large-scale, systemic changes.

5. **Adaptation plans do not sufficiently emphasize or incorporate investment considerations.**

Many countries are using their national adaptation plan (NAP) process to better coordinate and advance climate adaptation actions. A key milestone in the NAP process is the identification of national adaptation priorities, which are articulated in a NAP document or other strategic instruments such as nationally determined contributions (NDCs). As part of their NAP process, countries are increasingly conducting additional analyses, such as resource mobilization or financing strategies, to help transition from planning to implementing adaptation priorities. While these efforts provide valuable insights, they are often undertaken at an aggregate scale and prioritize international climate funds. Most countries still face persistent gaps in critical information that limit their ability to translate these priorities into investments capable of mobilizing other sources of finance (from MDBs), leveraging domestic public finance, and attracting the private sector (including through blended finance). In particular, the following information is often missing:

a. Robust climate, economic, fiscal, and financial justifications for investing in adaptation priorities, including evidence on the benefits of adaptation. This includes analysis of adaptation benefit streams and revenue-generation potential.

b. Detailed adaptation priorities organized into comprehensive, coherent investment packages that can attract finance at scale and build resilience across systems (cross-sector and cross-scale), with defined priority shortlists based on what is most urgent.

c. Targeted finance matchmaking that aligns adaptation investment packages to potential diverse funding and financing sources and various financial instruments.

6. **Adaptation investments are not reaching, and being managed at, the local level.**

The adverse impacts of climate change are predominantly experienced at the local level. However, estimates indicate that less than 17% of the total adaptation finance of about $95 billion allocated between 2017 and 2021 was directed toward local climate action (UNEP 2023: p. 45–46). Thus, to meet communities' adaptation needs and priorities, there is an urgent need to address this imbalance; for example, by strengthening countries' decentralized systems to channel climate adaptation finance to the local level.

Together, these challenges hinder developing countries' ability to scale up financing, which raises a critical question: How can developing countries mobilize public and private investments in adaptation more strategically—at the required speed and scale—to implement their national adaptation priorities and address climate risks?

3 APPROACH TO TACKLING GLOBAL CHALLENGES THROUGH ADAPTATION INVESTMENT PLANNING

Global initiatives on adaptation investment readiness

Globally, momentum is growing around strategies to address the challenges outlined in the previous section and to help developing countries unlock financing for adaptation, as shown by a wave of initiatives and frameworks launched in recent years.

COP30, held in Belem, Brazil, witnessed the launch of the NAP Implementation Alliance, developed as part of COP30 Action Agenda to accelerate collaboration and mobilize resources for implementing adaptation priorities. Table 1 summarizes examples of global and regional initiatives—primarily led by MDBs, climate funds, United Nations agencies, and programs supporting NDCs and NAPs—that aim to catalyze strategic investments in adaptation in developing countries.

Most of these initiatives were launched between 2022 and 2025, reflecting a heightened international focus on moving from adaptation planning to implementation. Given their recent establishment, most approaches are still in early stages of application, with notable exceptions such as support for the development of adaptation investment plans in various countries through initiatives led by the World Bank since 2018 and ADB since 2022.

A review of these initiatives reveals several common threads:

- Most approaches encourage countries to use adaptation priorities identified in NAPs and NDCs as entry points for unlocking finance, emphasizing alignment with national and subnational adaptation planning and budgeting systems.
- Common activities for preparing investment-ready adaptation priorities include prioritizing and sequencing investments, mapping adaptation priorities against diverse funding sources and appropriate financial instruments, and fostering an enabling environment for financing adaptation. Support for an enabling environment typically involves policy and regulatory alignment, capacity building, and engaging key actors.
- Approaches to prioritization often rely on cost–benefit analysis to assess the economic and financial return of adaptation investments. This reflects the methods currently used in most development financing. Some also use extended cost–benefit analysis methods to capture the characteristics of adaptation. For example, the World Bank uses probabilistic methods to assess the likelihood of various climate scenarios and their potential impacts, helping to reduce uncertainty around future climate impacts.

Table 1: Examples of Global and Regional Initiatives on Adaptation Investment Readiness

	Asian Development Bank (ADB)	NDC Partnership and the Green Climate Fund (GCF)	United Nations Development Programme (UNDP)	United Nations Disaster Risk Reduction, Standard Chartered Bank, KPMG International	World Bank	Institutional Investors Group on Climate Change (IIGCC)	Organisation for Economic Co-operation and Development (OECD)
Entity and initiative or guidance document	Climate adaptation investment planning program **(2022-** ***ongoing*****)**	Climate Investment Planning and Mobilization Framework **(2024)**	Adaptation finance strategy guideline **(2025)**	Guide for adaptation and resilience finance **(2024)**	Climate- smart agriculture investment planning **(CSAIP) (2018–*****ongoing*****)**	Physical Climate Risk Appraisal Methodology (PCRAM) **(2023)**	Climate Adaptation Investment Framework **(CAIF) (2024)**
Purpose	Translating national or sector priorities into actionable investments					Managing financial risks from physical climate impacts	Creating an enabling environment
Focus	Adaptation	Adaptation and mitigation	Adaptation	Sustainable development	Adaptation and mitigation	Adaptation	Adaptation
Scope	Sector, region, theme, or financing source	Multilevel	Sector	National level	Sector (agriculture)	Asset (e.g., infrastructure)	National
Output	Plan	Strategy/plan	Strategy	Strategy	Plan	Climate-proofed investments	Strategy

Source: Adapted by the authors from *CAIP Newsletter Issue 1 (2025)*.

The purpose of these initiatives varies. Many, such as ADB's Climate Adaptation Investment Planning program, the United Nations Development Programme's forthcoming guideline, the United Nations Department of Economic and Social Affairs' Integrated National Financing Frameworks, and the framework developed by the NDC Partnership and the Green Climate Fund (2024), focus broadly on translating national or sector adaptation priorities into investment pipelines. Others address a specific element of investment-readiness: for example, the Organisation for Economic Co-operation and Development (OECD) framework centers on creating a policy environment conducive to adaptation investment and is thus more focused on enabling conditions.

These new initiatives are critical as they draw attention to the growing gap between adaptation planning and implementation, and with coordination across initiatives, they can help developing countries receive coherent support from development partners.

The CAIP framework

In 2022, ADB in consultation with its developing member countries and development partners made progress with a framework, the Climate Adaptation Investment Planning (CAIP), designed to support the national scale-up of investment in adaptation (ADB 2024b). The framework, initially published in 2023 and refined based on experiences gained from its application, supports developing countries in translating their national adaptation priorities into investment-ready programs. It comprises five steps:

This step establishes the context for adaptation investment. It reviews the country's development and adaptation planning and financing frameworks and identifies relevant actors and entry points for adaptation investment. It establishes the sectors, regions, or programs for which the country aims to mobilize adaptation investments, ensuring they align with medium-term national and sector development planning processes. This step involves reviewing national and sector development plans, along with their associated priorities, objectives, and key performance indicators, as these determine investment and financing priorities. The goal is to ensure that the adaptation investments align with and strengthen government priorities. It also involves reviewing the country's public financial management processes, including its medium-term expenditure framework and public investment management criteria. This provides a fiscal context and informs the appraisal processes and prioritization criteria for public investments. Such alignment is essential for maximizing opportunities to mobilize financing from international and domestic sources. In combination, the review and alignment increase ownership of potential investments within ministries of finance and sector line ministries.

This step builds on existing climate risk assessments, such as those included in NAP documents or national communication, to develop more detailed information specific to a particular set of adaptation priorities to help inform adaptation investments. It focuses on understanding historic and current climate risks and considers future climate projections, taking account of uncertainty to analyze climate impacts on investment outcomes. The analysis adopts a more systemic approach, including multi-hazard analysis, and consideration of cross-sector risks. This step does not involve in-depth and complex modeling of climate impacts (something that can come later as part of detailed projects preparation) and instead identifies key indicators relevant for the sector of concern using 'ready reckoners' (proxies) to quantitatively assess the impact of climate change. For example, changes in the annual maximum 1-day rainfall total can be used as an indicator of flash floods. Information from this step is then used to help prioritize various adaptation investments.

This step considers the priority risks identified in Step 2 and groups individual adaptation activities into investment packages. These packages may address climate risks to a sector, region, theme or financing sources. It involves prioritization, moving away from long lists of adaptation actions toward specific, actionable investments. This prioritization is guided by adaptive management frameworks and considers the sequencing of options based on what is most important in the next 5 years (aligned with the medium-term planning process) to address both current and future risks. This is defined through a focus on (a) no-regret, low-regret, and win–win adaptation actions which provide immediate economic benefits, such as climate services for agriculture; (b) climate-smart design of long-lived investments or decisions that can be made in the near term, such as climate proofing new transport infrastructure; and (c) low-cost preparatory actions to improve future decisions, such as information gathering, research, piloting, and monitoring (Watkiss and Betts 2021). Ideally, adaptation measures are prioritized and grouped using participatory approaches and involving a broad range of actors, to ensure broad government ownership and a coordinated and cross-sector approach.

Step 4
Economic and Financial Rationale

This step assesses the costs and benefits of each adaptation investment package, including their potential economic and financial viability. This starts with an analysis of benefits, including the nature of the benefits (such as a reduction in crop yield losses or avoided flood impacts) and who is expected to receive the benefit. This analysis is critical for prioritizing and building support for the investment packages/activities and identifying the appropriate financing sources. First, the potential broad benefits of each adaptation investment package are assessed. Then, the gains are categorized from the perspective of different actors, looking at three sets: economic benefits (society and government), fiscal benefits (public investors), and financial benefits (private investors). This also helps identify the potential sources of finance in the next step.

This step includes an initial assessment of the economic internal rate of return (EIRR) and financial internal rate of return (FIRR) for each adaptation investment package. The economic analysis is consistent with standard appraisals by governments and MDBs, ensuring that proposed packages can meet thresholds set by these institutions or demonstrate net positive economic returns (as required, for example, for Green Climate Fund grants). This is complemented by an analysis of the FIRR and the financial sustainability of each option. While this information is relevant for governments and MDBs, it is particularly crucial for identifying revenue-generating projects that can attract private investors. More detailed economic and financial analyses will be required as the investment packages are developed further.

Step 5
Financial Matchmaking

This final step investigates the financing opportunities for each package, considering the multiple finance sources (public and private sources of domestic and international climate finance) and financial instruments (such as grants, guarantees, loans, debt, equity, and carbon credits) suitable for each package and tailored to specific country contexts, including fiscal space and access to capital markets. The analysis builds on the World Bank's decision tree for assessing financing sources and instruments (World Bank, Government of Nepal, and FAO 2021). The specific source of finance for each package is then developed by identifying institutions or programs that could be targeted. The analysis also considers whether policy reforms could be introduced to access diverse sources of finance or modalities and expand potential opportunities.

Cross-cutting element and next steps. Across the five steps, the CAIP framework also acknowledges the vital importance of a supportive enabling environment—including strong government ownership, institutions, and capacity, and improved awareness and capacity among domestic finance institutions and private sector actors. This is complemented by the analysis of enabling factors that could unlock finance, for example, through regulatory or policy reform. This leads, in turn, to a set of recommendations that can advance next steps. Common next steps include the launch of the CAIPs by governments, the organization of development partner and investor forums, and the development of concept notes for priority investment packages for submission to secure domestic budget and/or to potential financial institutions and funds.

By applying these steps, the CAIP framework offers a novel approach to addressing key global challenges in mobilizing climate adaptation investments, as summarized in Table 2. Further, Box 2 explains how the CAIP framework strengthens the NAP process.

Table 2: How the CAIP Framework Addresses Global Challenges in Mobilizing Climate Adaptation Investments

Global challenges in mobilizing climate adaptation investments	CAIP framework solutions
1. Climate risks will likely continue to intensify and threaten development progress.	The climate adaptation investment process begins with a country's development and adaptation goals for the relevant sector and region, rather than starting with climate change projections. This approach increases coherence through policy mainstreaming and maximizes impact by ensuring that financing supports not only climate resilience within the selected sector but also advances broader development objectives. Adaptation investment packages are identified to address multiple climate risks at a system level, allowing different actors to work together to prioritize comprehensive, cross-sector solutions. For example, commercial banks are unlikely to invest in climate-resilient agricultural production without accompanying actions on processing and market potential. This is why a CAIP for the agriculture sector identifies adaptation investment packages along value chains.
2. The speed and scale of adaptation financing and implementation remain insufficient.	Speed is addressed by aligning with and strengthening national development and adaptation priorities, resources, and processes; identifying the most significant present and future climate risks to the intended development outcomes of near-term adaptation investments in the sector or area of focus; and by prioritizing the most urgent investments. Scale is addressed by moving from projects to a programmatic approach and by identifying multiple sources of finance and a variety of financial instruments (matchmaking) to deliver this. This diversifies and thus increases financing by leveraging domestic resources and mobilizing private sector investment alongside international support.
3. Understanding of the fiscal risks of climate change and adaptation is limited.	The CAIP process engages the Ministry of Finance (or its equivalent) as a primary actor from the outset. The analysis aligns with and integrates into government public financial and public investment management processes, and it identifies the fiscal benefits of adaptation packages. The finance matchmaking step also examines fiscal instruments the country can use to expand domestic resources for adaptation.
4. Programmatic approaches to adaptation are lacking.	The approach begins with adaptation priorities identified in strategic documents—such as the NAPs or NDCs. These priorities are further developed to unlock finance at scale for a priority sector or area, moving beyond individual projects and working with investment packages to deliver within different programs. It also incorporates climate risk diagnostics at the system level to design programmatic adaptation investment packages that address multiple climate risks and cross-sector linkages. This process encourages the involvement of various ministries to ensure alignment of the adaptation investment plan with national development and adaptation priorities—and so supports national ownership and long-term sustainability.
5. Adaptation plans do not sufficiently emphasize or incorporate investment considerations.	The approach addresses key information gaps identified in NAPs and other strategic documents, particularly: (a) assessing the benefits of investing in adaptation—whether these are economic (societal) or involve potential financial returns—and identifying possible sources of finance, including public, private, and blended finance; (b) assessing the potential economic and financial returns of adaptation packages and providing an investment rationale for public and private actors; (c) conducting targeted finance matchmaking to align diverse sources of finance and financial instruments with each investment package and subpackage; (d) extending the analysis to identify specific funding and financing opportunities; and (e) assessing potential regulatory or policy reforms that could address barriers and help scale up finance.
6. Adaptation investments are not reaching, and being managed at, the local level.	The approach identifies various channels through which funds can reach the local level, such as national banks, and explores opportunities for local private sector actors to invest in adaptation priorities. Where relevant, it engages with decentralized government entities, such as provincial governments, to directly identify locally specific adaptation investment packages and embed them within local governance structures and processes.

Note: CAIP = Climate Adaptation Investment Planning; NAP = national adaptation plan; NDC = nationally determined contribution.
Source: Authors.

Box 2: How the Climate Adaptation Investment Planning Framework Strengthens the National Adaptation Plan Process

The national adaptation plan (NAP) process was formally established in 2010 under the Cancun Adaptation Framework at the 16th Conference of the Parties (COP16) of the United Nations Framework Convention on Climate Change (UNFCCC 2011a). Its objectives are to "reduce vulnerability to the impacts of climate change, by building adaptive capacity and resilience, and facilitate the integration of climate change adaptation, in a coherent manner, into relevant new and existing policies, programmes and activities, in particular development planning processes and strategies, within all relevant sectors and at different levels, as appropriate" (UNFCCC 2011b).

Simply put, the NAP is a country-driven process that seeks to systematically integrate adaptation considerations into development decision-making, rather than treating adaptation as a stand-alone exercise (Hammill, Dazé, and Dekens 2020). It builds systems and capacities for governments and other actors to address medium- and long-term climate risks through ongoing, iterative cycles of risks assessment, planning, implementation, monitoring, evaluation, and learning—ultimately aiming for climate-resilient development.

A key milestone in the NAP process is the identification of national adaptation priorities, which are articulated in a NAP document or other strategic instruments such as nationally determined contributions (NDCs).

Importantly, the NAP process does not end with identifying national adaptation priorities; it is a continuous mechanism for identifying, coordinating, and addressing a country's adaptation needs in a more coherent and strategic manner. Therefore, mobilizing financial resources and implementing adaptation priorities are integral parts of the NAP process.

Beyond identifying their national adaptation priorities, countries are therefore increasingly conducting additional analyses, such as resource mobilization or financing strategies, to help transition from planning to implementing their priorities. While these efforts provide valuable insights, they are often undertaken at an aggregate scale and prioritize international climate funds.

Most countries still face persistent gaps in critical information that limits their ability to translate these priorities into investments capable of mobilizing other sources of finance (such as from the multilateral development banks), leveraging domestic public finance, and attracting the private sector (including through blended finance). This means that while NAPs and related strategic documents often serve as entry points for prioritizing adaptation investments, investment readiness is not yet explicitly embedded as a core element of the NAP process.

At COP30 in 2025, countries adopted the NAP assessment decision, recognizing developing countries' progress in adaptation planning and implementation while noting ongoing challenges (UNFCCC 2025c). Although the decision affirms the role of NAPs in global climate architecture, it provides limited guidance on scaling up support for developing countries' NAP processes.

Across the six countries where the Climate Adaptation Investment Planning (CAIP) Framework was applied, insights from discussions with key actors revealed that:

- Adaptation has not been fully integrated into development planning processes that enable the design of adaptation investment packages linked to public and private financing models.
- NAPs often lack visibility and ownership across ministries and are perceived as static documents rather than iterative processes essential for achieving climate-resilient development.
- Adaptation is still viewed as an additional or competing priority, rather than as a fundamental requirement for meeting development goals.

The **Framework** aims to address these gaps by supporting decision-makers to take an investment-led approach to:

- **Strengthen system-level risk analytics** to design programmatic adaptation investment packages.
- **Conduct economic analysis** to demonstrate the return on investment in adaptation and allocate responsibilities across the public and private sectors.
- **Facilitate finance matchmaking** to leverage public and private financing for adaptation investments.
- **Enhance the policy and regulatory environment** to enable and sustain adaptation investments.

The Framework is guided by public financial management systems, ensuring that adaptation investments are embedded in development planning and budgeting processes. It has been identified in the Baku to Belem Roadmap as a guide to help implement the NAP process by supporting the integration of adaptation into development planning (UNFCCC 2025a).

Source: Authors.

Overview of key results from piloting CAIP in six countries

Table 3 summarizes the results of applying the CAIP framework in six countries: Armenia, Bhutan, Cambodia, Indonesia, Mongolia, and Nepal.

Table 3: CAIPs at a Glance

Country	Goal of CAIP and lead ministry	Main entry point(s) for identifying adaptation investment in the sector or area of focus	Adaptation investment packages
Armenia	Mobilize investment in climate-smart agriculture, followed by a focused deep dive on climate-resilient horticultural value-chains *Lead: Ministry of Economy's Agriculture Department, with involvement of the Ministry of Finance and Ministry of Environment*	• Updated NDC (2021) and NAP document (2021) • Agriculture Development Strategy 2020–2030 (2023), which integrates the Agriculture Sectoral Adaptation Plan	Seven investment packages designed to address climate change impacts across the entire horticulture value chain—from production through processing to marketing *Estimated funding needed: Between $150 million and $190 million by 2050 (adaptation needs in agriculture)*
Bhutan	Mobilize investment in adaptation in the water sector *Lead: A cross-government Inter-Agency Working Group led and coordinated by the Department of Water, Ministry of Energy and Natural Resources and Ministry of Finance*	• NAP water sector road map (2024), which draws from Bhutan's first NAP document (2023) • 13th 5-year national development plan (2024–2029)	Eleven investment packages covering resilience aspects related to water supply (water shed management, climate resilient water supply), water demand (irrigation, hydropower), water induced disasters (flood risk management, urban planning and green infrastructure, disaster management and emergency response), and cross-cutting themes such as integrated water resources management, adaptation policy and planning, research and education, national climate services and early warning systems *Estimated funding needed: ~$850 million over a 15–20-year period*
Cambodia	Mobilize investment in climate-resilient agriculture (rice, fish, high-value export crops) for a special economic zone (comprising four provinces in the northwest) *Lead: Ministry of Economy and Finance with involvement of Technical Advisory Group composed of line ministries*	• Draft Climate Change Priority Action Plan for the Agriculture Sector (2030), based on Cambodia's second NAP document, the Cambodia Climate Change Strategic Plan II (2024–2033) • 5-year national development plan	Seven investment packages covering planning, crop production, transport and logistics, processing and storage, and cross-cutting themes such as weather and climate services, integrated land management, and integrated water resources management *Estimated funding needed: $500 million*

continued on next page

Table 3 continued

Country	Goal of CAIP and lead ministry	Main entry point(s) for identifying adaptation investment in the sector or area of focus	Adaptation investment packages
Indonesia	Mobilize investment in climate-smart agriculture, followed by a focused deep dive on a financing scheme, the Green Kredit Usaha Rakya for climate-smart agriculture, for micro, small, and medium-sized enterprises *Lead: Ministry of Finance with involvement of Ministry of Agriculture, Financial Services Authority, and Ministry for Economic Affairs*	• National Medium-Term Development Plan 2020–2024 • Low-carbon and climate-resilient road map (draft) • Updated NDC (2022) • Climate Resilience Development Policy • Sustainable Finance Taxonomy • Green Bond and *sukuk* frameworks	Three investment packages focused on climate-smart agriculture models, financial models, and policy and institutional capacity strengthening to support farmers investing in climate-smart agriculture *Estimated funding needed: $544.4 million to $2.18 billion to extend interest subsidies over 1–4 years for 4.75 million hectares of irrigated rice fields*
Mongolia	Mobilize investment in climate resilient water sector, followed by a focused deep dive on climate-resilient water ponds in different regions *Lead: State Water Agency, Ministry of Environment and Climate Change*	• NAP document (2025) • NDC 3.0 (2025) • Vision 2050's Goal 6 on Green Development (2020) • Relevant water and agriculture sector policies	Three investment packages focused on (1) Hydromet monitoring and land-use planning services; (2) Project preparation facility and (3) Climate-smart water pond program to build, manage and maintain 250 climate-smart water ponds over a 5-year period, through to 2030 *Estimated funding needed: $56.15 million over 5 years*
Nepal	Secure investment in climate resilient agriculture (rice and horticulture) in Madhesh province Lead: Ministry of Agriculture and Livestock Development with involvement from the provincial Ministry of Land Management, Agriculture and Cooperative	• Adaptation priorities for agriculture identified in Nepal's first NAP document (2021) and updated NDC (2025) • Agricultural development strategy • National Strategic Development Plan (the 16th 5-year plan) • Madhesh second periodic plan • Madhesh agricultural master plan	Seven investment packages covering agriculture value chains, and cross-cutting themes including as weather and climate services, and integrated water resources management *Estimated funding needed:* $360 million to $700 million over the next 5 years

Note: CAIP = Climate Adaptation Investment Planning; NAP = national adaptation plan; NDC = nationally determined contribution.
Source: ADB. 2025a, 2025b; ADB, forthcoming-d, -e, -f, and -g.

In each country, activities and results from applying the CAIP framework were summarized in individual reports.

Key results are detailed in three technical briefs (ADB, forthcoming-a; ADB, forthcoming-b; ADB, forthcoming-c). With a focus on the economic and finance side, these can be summarized as follows:

- Nearly all adaptation investment packages in the agriculture and water sectors show high net-positive economic benefits, justifying their public prioritization. The analysis of packages indicated high economic internal rates of return that would pass thresholds in Government Public Investment Management criteria, and similarly the thresholds for approval by MDBs (for example, above the 9% threshold used by ADB).
- Financial returns from many of the adaptation packages are significantly lower than economic returns, highlighting challenges to financial viability. This reflects a greater focus on public sector priorities in NAPs, and these are likely to require public finance. However, some packages also demonstrate financial returns, and these can attract potential private investment.
- Increasing private adaptation finance is sector- and package-dependent. The early opportunities are in areas where the private sector is already active, such as in agricultural value chains in countries like Armenia, Cambodia, Indonesia, and Nepal. Government actors recognized their critical role in creating incentives for private investment in adaptation and identified potential levers for de-risking investments including

addressing information and awareness gaps; advancing policy and regulatory reforms; and using blended finance instruments such as grants, guarantees, and concessional credits.

- Opportunities were identified to increase private investment in other investment priorities over the longer term. This includes areas with potential revenue, such as agricultural research and development or the provision of value-added climate services to farmers. It can also build on other initiatives that may be seeking to encourage more private investment—e.g., where there are plans for greater private involvement in the water sector such as through public–private partnerships. These are conditional on the necessary political economy and the willingness of governments around broader regulatory reform.

- There is not one single source of finance (or financial instrument) that can operationalize an adaptation investment plan. Many adaptation investment packages comprise a blend of interventions including institutional, infrastructure, and technology-based adaptation solutions. Often, financing needs to support both capital investment and recurrent expenditure as well as encourage reforms. This can be achieved through project financing with complementary technical assistance or financing packages that combine investment loans with policy-based financing.

The next section presents key insights from the process of applying the CAIP framework in the six countries.

Climate-smart water ponds in Mongolia can strengthen the resilience of rangelands and livestock.© Nanki Kaur

4 LESSONS FROM APPLYING THE CAIP FRAMEWORK IN SIX COUNTRIES

This section presents key lessons from applying the CAIP framework in the six countries.

Adaptation investment planning must operate within the existing development planning and fiscal context of the country, requiring a careful balance between ambition and pragmatism.

Each country prioritized adaptation investments that were feasible within their existing development planning and fiscal contexts. This alignment not only fostered buy-in and legitimacy but also necessitated balancing ambition with pragmatism.

Most developing countries encounter a range of persistent development barriers that hinder their ability to mobilize investments for adaptation in a more coordinated and effective way. These include misalignment between planning and budgeting, limited subnational capacity to identify investment opportunities, significant fiscal deficit, sector-based silos, and frequent turnover of government staff. Aligning the preparation of investment-ready adaptation priorities with these realities was important to ensure a pragmatic approach that balanced ambition with the realities of national development planning and budgeting systems.

Take Nepal as an example. The decentralization process has been ongoing since the adoption of its 2015 Constitution, which established a federal system and assigned to provinces the responsibility for development planning and implementing programs in their regions. However, preparing investment-ready adaptation priorities in different provinces remains challenging because local governments, who depend on the federal government to borrow, face limited financial capacity and constraints. Decentralization has also introduced additional complexity by increasing the number of actors, which complicates coordination (Pokhrel 2024). For instance, planning commissions and sector ministries such as agriculture now exist at both federal and provincial levels. As a result, development planning has become more difficult to harmonize. For example, Nepal's federal 16th 5-Year Development Plan is organized by thematic areas, whereas the 5-year plan of Madhesh Province (which was selected by the government for applying the CAIP framework) is arranged by sector, illustrating differences in planning approaches across different levels of governance. In addition, there is an important municipal level government (over 700 municipalities) which is responsible for implementing most of developmental interventions of the federal government as well as its own plans, budget and taxes.

In Mongolia, it was crucial to align the timing of the adaptation investment planning process with the country's public investment programing cycle to ensure adaptation priorities were integrated into the 5-year investment pipeline and linked to budget frameworks.

In Armenia, the government showed strong interest in pursuing the CAIP horticulture program. To support this, a detailed economic and financial return analysis was completed and used to complete the government template for Public Investment Management and thus support prioritization of the package.

Beyond the need to find a middle ground between ambition and pragmatism, countries had to manage other trade-offs in preparing investment-ready adaptation priorities, calling for strong government steer and coordination. In Cambodia, for example, the process highlighted the need for strengthening alignment between government priorities—primarily focused on developing rice and fish value chains—and investor priorities, interested in climate-resilient commercial crops.

A focused and iterative approach is essential for adaptation investment planning.

The approach to applying the CAIP framework was both focused and iterative.

First, on the focused approach: Given the complexity and details required, simultaneously preparing every adaptation priority outlined in a national adaptation plan (NAP) document or other strategic documents for investment was found to be both impractical and ineffective. To mobilize financial resources, governments need to access context-specific information on adaptation priorities—detailing what will be done, by whom, where, how, and over what period. Governments therefore had to adopt a focused approach to adaptation investment planning. This was done by selecting a manageable unit of focus, which could be a sector (e.g., water), theme (e.g., value chain), or specific subnational region or economic zone, depending on national priorities and governance structures.

Narrowing the scope of the adaptation investment plan (for example, in Bhutan, the government focused on the water sector for the CAIP) not only made it more manageable to prepare investment-ready adaptation priorities but also increased their attractiveness to potential public and private financial actors by producing well-defined, investment-ready projects.

Second, on the iterative approach: countries adopted an iterative and incremental approach to preparing investment-ready adaptation priorities. This flexible method allowed investment packages to be continuously refined.

Countries conducted multiple rounds of adaptation option prioritization, using feedback and new information to focus efforts on the most promising and feasible investment opportunities. In particular, the prioritization of adaptation investments occurred when selecting the sector, region, or theme of focus, then additional prioritization was based on the results from more granular climate risk analysis and the economic and financial appraisal. Carefully sequencing investments—by identifying which interventions should be undertaken first to lay the groundwork for subsequent actions—helped ensure the coherence of adaptation priorities and address any gaps. In some cases, results from the finance matchmaking exercise also served as a further review of the adaptation investment packages to ensure a diversity of funding sources.

Tailoring climate risk diagnostics to specific adaptation priorities helps prioritize and justify investments.

Understanding climate risks is fundamental to effective adaptation, and all countries had already conducted some form of climate risk assessment to identify their adaptation priorities. Climate justifications also influence potential financing sources and determine the concessional financial support required. These justifications enable governments to articulate their needs in ways that align with investor expectations. From an investor perspective, strong climate justifications help build confidence.

All six countries required more focused and up-to-date analyses of climate risks linked to the adaptation priorities for the selected sector, subsector, or region and considering overlooked aspects such as multi-hazard analysis, cross-sector risk linkages, and uncertainties. These findings are consistent with recent studies. For instance, the United Nations Environment Programme highlights gaps in the quality of climate information integrated into NAP documents (UNEP 2024), while the NAP Global Network underscores the need for better alignment between climate risk assessment results and the adaptation measures prioritized within NAP documents (Dekens 2023).

The climate risk diagnostics therefore complemented existing information and provided more specific data needed to further prioritize and detail particular adaptation priorities.

Strengthening the climate evidence base involved reviewing adaptation actions against latest climate science. In the case of Bhutan, this was achieved by addressing previously overlooked flood risks from glacial melt in its NAP document. It also involved conducting more granular climate analysis to fill specific gaps and avoid maladaptation. For example, Cambodia developed storylines for chronic heat risks and precipitation variability affecting agricultural productivity and cold chains. Four storylines (Figure 1) were developed for northwest Cambodia to reflect changes in global mean temperature (T) and regional climate variability (V) during El Niño and La Niña events by the 2050s: warm-wet (Ww), warm-dry (Wd), hot-wet (Hw), and hot-dry (Hd). El Niño+ and La Niña+ represent the expected changes in annual mean temperature (ΔT) and precipitation (ΔP) under higher global warming. The results help decision-makers consider a wider range of climate conditions (ADB, forthcoming-c). Results also revealed heat risks to labor productivity—which had been omitted from the agriculture sector NAP document and was subsequently added to an adaptation investment package.

Mongolia conducted scenario analyses of pond water volumes under current and projected climate conditions at three test sites to determine optimal pond design and location in a changing climate. The results showed that available water depends on factors such as the initial pond volume at the start of summer, the timing of summer rainfall and runoff, spring flow levels, and sedimentation rates. A checklist for climate-smart pond siting, design, and management was developed to help minimize risks of sedimentation, water shortage, and flood damage.

Figure 1: Four Climate Change Storylines for Northwest Cambodia

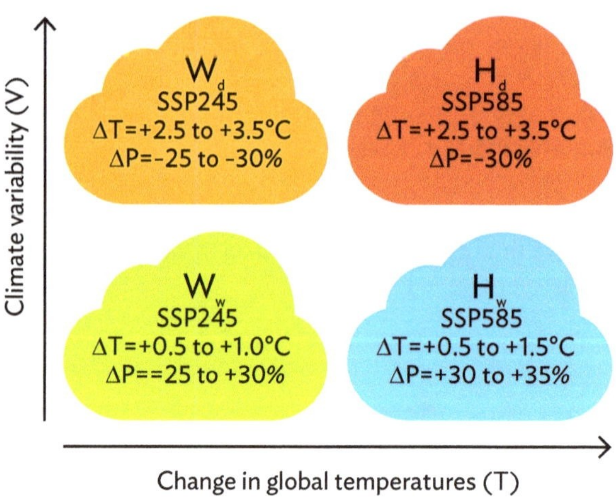

Note: SSP = Shared Socioeconomic Pathway.
Source: ADB, forthcoming-c.

Comprehensive adaptation investment packages address multiple climate risks at a system level and include different types of investments.

While NAP and other strategic documents provided foundational direction, the identification and prioritization of adaptation investment packages often required refinement to identify specific sources of financing and funding. For example, many NAP documents list broad priorities like "improve water security" and "climate-smart agriculture", and/or identify specific technologies. However, identifying suitable financing sources and instruments requires details on adaptation priorities (such as exact locations and infrastructure needs) as well as consideration of implementation modalities. For example, in climate-smart agriculture, details are needed on whether finance supports extension services, subsidies, and/or technology access.

In addition to detailing adaptation priorities for a selected sector, region, or theme, countries' experiences demonstrated the necessity and feasibility of grouping these priorities into comprehensive (or programmatic) adaptation investment packages. Grouping adaptation priorities into packages was important to support "system transformation", moving beyond the traditional focus on individual projects. These packages enabled each country to assess whether the selected sector would be fundamentally transformed under various climate scenarios. In this context, comprehensiveness meant designing investment packages that addressed five core requirements:

- Tackling multiple instead of single climate risks while accounting for uncertainties associated with future climate scenarios. This approach encouraged the diversification of adaptation options—for example, in Cambodia, it promoted alternatives to rice production by introducing a new focus on horticultural and root crops, which are less sensitive to extreme heat and precipitation variability than rice.

- Building the climate resilience of entire systems. Effective adaptation requires a systems-based approach, one that recognizes the complex relationships and interactions among various actors, whether within a country, region, ecosystem, sector, or value chain. This focused on cross-sector and cross-scale solutions to help prevent maladaptation by ensuring that interventions consider the broader context. For example, adaptation investments in the Bhutan water sector considered different packages addressing water demand, water supply, and water-induced disasters. In Armenia, Cambodia, and Indonesia, investments in agriculture considered packages addressing different impacts and needs along value chains, from production to marketing. Mongolia's climate-smart ponds investment program aimed to enhance rural water supplies, grazing management, small-scale vegetable growing, and irrigated fodder planting. Importantly, as noted earlier, investors need information on investment benefits at the system level to commit funds.

- Covering different types of adaptation investments that determine access to different financial instruments including concessional finance. For example, in Bhutan, adaptation packages identified for climate-resilient national water sector development include four types of investment: (i) climate-proofing investments for water sector development (such as climate proofing of new sewerage systems); (ii) investments in water sector development with a partial adaptation focus (such as integrating climate risk into integrated water resources management analysis and implementation); (iii) investments in climate mitigation or water sector development with potential adaptation co-benefits (such as investments in forestry with potential adaptation co-benefits from upstream catchment management through reduced runoff and flow regulation); and (iv) dedicated climate adaptation investments to address specific vulnerabilities in the water sector (such as investment in climate services and early warning systems).

- Including investments aimed at strengthening the enabling environment, which is essential for de-risking other investments and ensuring the sustainability and quality of financing. The enabling environment encompasses appropriate policies and regulations, incentive mechanisms, institutional arrangements, access to data and information, and the development of necessary skills and capacities. A thorough understanding of this environment is vital for identifying and removing barriers that could undermine adaptation investments. Bhutan provides a practical example, having developed adaptation investment packages that integrate adaptation considerations into integrated water resources management, strengthen national climate services, and enhance monitoring and evaluation systems for transparency and accountability.

- Including packages of significant size. Structuring adaptation investments at a scale substantial enough to attract larger flows of private and international public finance is crucial. Of course, what constitutes a "significant size" depends on the context.

Taken together, the adaptation investment packages were designed to address adaptation in a comprehensive and integrated way—in contrast to individual projects, which often tackle adaptation in a more fragmented and isolated manner.

Appraising the benefits and viability of adaptation investment packages can facilitate their integration into public management processes and identify opportunities for the private sector.

Countries are increasingly quantifying adaptation costs through their NAP processes. For example, 37 of 59 NAP documents submitted to the United Nations Framework Convention on Climate Change as of February 2025 include cost estimates often disaggregated by specific actions, sectors, or subnational units (Hernández et al. 2025) and many NDCs also cost adaptation priorities. However, all lack quantified analysis of the benefits—let alone the economic and financial returns—an essential gap for attracting adaptation investment, as highlighted by Brandon et al. (2025).

Results from applying the CAIP framework showed that it was feasible and useful to conduct qualitative and semi-quantitative assessments of the benefits to build the case for investing in adaptation. This included a mapping of what the adaptation benefits are, and who benefits. Detailed quantitative appraisals will still need to be completed later during project preparation.

At this early stage in development of the packages, a detailed economic and financial analysis was not possible, particularly for investment plans targeting many packages, due to data requirements. However, a light-touch appraisal of the benefits was complemented with a review of the potential economic and financial returns, based on the literature and on similar projects from development partners. This provided key information on the attractiveness of the investments and, in turn, whether they will meet Public Investment Management criteria commonly used in government prioritization and economic criteria applied by MDBs in financing, as well as on whether they offer a sufficient financial return for private investment. For example, ADB has a minimum economic internal rate of return (EIRR) of 9% for investment (ADB 2017).

As examples, analysis in Cambodia and Nepal found that all the seven agricultural packages were likely to deliver a high EIRR that would exceed government Public Investment Management and ADB thresholds, with all packages exceeding a 10% EIRR and in many cases returns several times higher than this. This provides key evidence that of the economic return, and justification.

As packages progress, more detailed cost–benefit analysis appraisal can be undertaken. For example, in Armenia and Indonesia, deep dives were undertaken on individual packages, focusing on climate-smart horticulture in Armenia and a green line of credit for the Kredit Usaha Rakya in Indonesia. For both, a detailed economic and financial appraisal was undertaken to provide quantitative analysis of public and private returns.

Decision trees serve as a practical starting point for matching adaptation priorities with financing sources and instruments, guiding countries toward actionable steps.

In addition to improving understanding of each adaptation investment packages and their revenue-generating potential, countries also developed a more detailed knowledge of the potential finance providers, the instruments they use, their associated terms, and the barriers they may face. Considering these issues is a crucial step in encouraging actors to think more concretely about implementation. Experience showed that when key actors began considering financing, it helped shift the work from abstract concepts to practical steps.

While some countries had identified broad financing sources for their national adaptation priorities, more granular matchmaking was essential to align refined adaptation investment packages with appropriate sources of finance (public, private, blended) and financial instruments, such as grants, debt, and guarantees. This process required context-specific alignment with government priorities and the requirements of potential finance providers to improve coordination and coherence. It helped clarify the entities most likely to finance (cover up-front costs), fund (repay the financing costs over the lifetime of the project or asset), implement, or support specific adaptation investments.

Experience from the countries demonstrated that a decision tree was a useful starting point to begin this process. This approach, which built on the World Bank's 2021 decision tree for climate-smart agriculture financing for Nepal (World Bank, Government of Nepal, and FAO 2021), was applied in all six countries and is illustrated in Figure 2 (ADB, forthcoming-b). Compared to the decision tree of the World Bank, Nepal, and the Food and Agriculture Organization of the United Nations, this framework provides greater detail on financial instruments (the right column of Figure 2).

The first step in the decision tree is to ask whether the investment package is currently financially viable and if private sector finance is already involved. The choice of financial instruments then depends on the investment size and risk profile, with public sector roles focused on supporting and scaling through championing leading companies.

If the private sector is not already financing the activity, the next step is to evaluate whether policy and regulatory reforms could enable private investment, contingent on a strong government commitment. Such reforms can be supported by development partners through technical assistance or policy-based lending from international financial institutions, which provide budget support conditional on completed policy reforms. If the enabling environment reforms are insufficient then it considers whether different forms of blended finance instrument may help leverage private sector investment, with the appropriate instrument varying according to the barriers that might be holding back private investment. Finally, there are a range of adaptation activities for which only public finance will be suitable, with the appropriate financial instruments to deliver this public investment broadly depending on whether it is an investment or ongoing costs, the extent to which the primary objective of the activity is adaptation, and the country in which the adaptation is taking place.

Figure 2: Decision Tree for Assessing Financing Sources and Instruments

Greater private sector delivery* ↑ … ↓ Greater public/civil society sector delivery

Economic and financial characteristics of asset/activity		Government role	Illustrative actions
A. Is the private sector* already doing it? (No ↓)	*Implies FIRR > private sector hurdle rate* (Yes →)	Support and scale	• Improve enabling environment • Convene additional players • Champion and platform • Support MEL
B. Based on assessment of likely cashflows and international experience, could the private sector do the activity with changes in policies and regulations (only) and is this desired? (No ↓)	*Implies FIRR with policy/regulatory changes > private sector hurdle rate* (Yes →)	Improve enabling environment	• Address policy/regulatory gaps • Address moral hazard—clarify assumptions around availability and role of public and private sector • Support the development of new business models
C. Can public finance crowd in private investment through changing risk-return profile and is this desired? (No ↓)	*Implies EIRR > threshold but FIRR > private sector hurdle rate only after the application of blended finance* (Yes →)	Use blended finance and de-risking	• Develop blended finance approaches • Improve governance and coordination amongst and across public and private sectors • Support knowledge sharing and information • Revise public funding terms and conditions over time to encourage leveraging/de-risking
D. Do cashflows characteristics—or choices about how the activity is organized and whether assets should be owned by the public sector—mean that the activity/asset will be publicly financed into the medium term?	*Implies EIRR > threshold but FIRR < private sector hurdle rate, or that other factors lead to preference for public sector provision* (Yes →)	Boost spectrum of public financing	• Identify timing and suitability of existing public funding sources • Mainstream adaptation requirements into line ministries, government agencies and sector programs • Use green budgeting and climate fiscal planning to understand the value of adaptation and assess progress

* It may also be possible for corporatized state-owned enterprises (SOEs) to undertake the same activities as the private sector. These bodies could access various debt instruments but some/all ownership would reside in the public sector, meaning that they could not access private equity finance.
The pros and cons of SOE versus full private ownership will depend on national circumstances. SOE models may promote public accountability and better serve nationally determined development objectives but may risk greater operational inefficiency and can create a fiscal burden if they make losses.

Potential financial instruments and key considerations

Scale of investment	Cash flow risk: Low ⟷ High		
Low ↑ … High ↓	Microfinance loans Local Financial Institution loans International lending Bonds**	Equity	Venture funding

Development–adaptation continuum

	Significant proportion of activity's prospective cashflows not linked to adaptation/public intervention justified by conventional development priorities ⟷	Almost all activity's prospective cashflows associated with building climate resilience/adaptation
All countries	National and local government commitment to enabling environmental reform and Technical assistance and policy-based support Conventional development partner	Dedicated climate finance

Development–adaptation continuum

Key barrier to private investment	Significant proportion of activity's prospective cashflows not linked to adaptation/public intervention justified by conventional development priorities ⟷	Almost all activity's prospective cashflows associated with building climate resilience/adaptation
Project economies	*Grants, subsidies (including performance-based grants)* Government — Development partners	Dedicated climate finance
Limited investment finance	*Concessional debt, patient equity (according to risk profile activities)* Development partners	Dedicated climate finance
High risk profile	*Guarantees, insurance, subordinated debt and equity* Development partners	Dedicated climate finance
Skills/capacity	*Project development support* Development partners	Dedicated climate finance

Country status	**Actions lead to ongoing costs** **Development–adaptation continuum** More aligned to existing development priorities	Primary objective is adaptation	**Actions lead to ongoing costs** **Development–adaptation continuum** More aligned to existing development priorities	Primary objective is adaptation
Middle-income with more fiscal space and capital market development, lower climate vulnerability	*Taxes/budget reallocation*	*Budget support/grant*	*Commercial loans/(green) bonds* *Conventional IFI loans*	*Concessional climate finance*
Low-income and/or limited fiscal space and/or acutely climate vulnerable	*Taxes/budget reallocation*	*Budget support/grant*	*Conventional IFI loans*	*Concessional climate finance or debt resolution activities*

** Subject to international market access or sufficient local capital market development.

Note: EIRR = economic internal rate of return, FIRR = financial internal rate of return, IFI = international financial institution, MEL = monitoring, evaluation, and learning.

Source: ADB, forthcoming-b.

The decision tree thus served as a starting point for identifying appropriate financing instruments for specific adaptation investment packages. However, more detailed analysis was still needed to pinpoint relevant financing sources and funding mechanisms for each country context—answering questions such as which providers offer those instruments for the relevant investment packages. Results showed that combining literature reviews, interviews with government representatives and staff from MDBs, and roundtables with development partners enabled such detailed analyses.

Comprehensive adaptation investment packages and targeted finance matchmaking require broadening and deepening engagement with key actors.

Broadening and deepening the engagement of key actors was critical for building consensus, ensuring ownership, identifying comprehensive adaptation investment packages, and identifying suitable financing sources and instruments. The terms "broadening" and "deepening" recognize that most countries had already begun expanding and intensifying their engagement on adaptation with key actors as part of their NAP processes—efforts that needed to be strengthened and sustained. In each country, more comprehensive actor mapping to identify relevant ministries, agencies, public and private financial institutions, and other actors was conducted for the chosen sector, theme, or region.

Within government, coordination among ministries responsible for development planning, finance, and climate adaptation was crucial for effective preparation of investment-ready adaptation priorities. Traditionally, environment ministries have led NAP processes, with limited involvement from ministries of finance or development planning. Yet, ministries of finance have a key role in ensuring that adaptation investments align with a country's economic strategy, fiscal policies (including taxation, debt, and budget management), financial policy, and overall financial system, as promoted by global initiatives such as the Coalition of Finance Ministers for Climate Action (Coalition of Finance Ministers for Climate Action 2023). Where domestic finance is involved, ministries of finance are pivotal to integrating adaptation priorities into national budgets to ensure their implementation.

Across the six countries, investment-readiness activities gave an opportunity to strengthen the engagement of finance ministries in national adaptation efforts. For example, in Bhutan, the Ministry of Finance played a key role in initiating the CAIP process by leveraging an ADB policy-based loan, which included adaptation targets identified in the country's NAP document. In Cambodia, the Ministry of Economy and Finance was engaged as a co-implementer of the adaptation investment planning process from the outset and through regular meetings, which contributed to strengthened coordination among, and participation from, line ministries.

Engagement with finance ministries in the CAIP process is essential to enhance the potential for mobilizing financing for adaptation from diverse sources. Crucially, this engagement must be closely coordinated with the ministries responsible for the NAP process to support a coordinated approach to transitioning from planning to implementation. This collaboration facilitates learning for preparing investment-ready adaptation priorities, replication of the approach across sectors or regions, and ensuring that the process and outcomes inform subsequent NAP updates.

The involvement of sector ministries was tailored to the CAIP's focus, especially as programmatic approaches to adaptation increasingly required integrated, cross-sector solutions to ensure coherence, avoid maladaptation, and maximize impact. Preparing adaptation priorities for investment-readiness was therefore in most cases an opportunity to strengthen cross-sector collaboration and solutions. In Bhutan, for example, the preparation of investment-ready adaptation priorities for national water development led to unprecedented collaboration among all ministries involved in the sector. While the Department of Water under the Ministry of Energy and Natural

Resources holds overall responsibility for water management, other ministries play key roles: the Department of Energy oversees hydropower generation; the Ministry of Infrastructure and Transport manages water infrastructure projects and flood risk management; the Ministry of Agriculture and Livestock is responsible for irrigation in agricultural development; and the Ministry of Finance mobilizes resources for climate-related activities through mechanisms such as the Bhutan Climate Fund.

Engagement extended beyond government to the entire financial ecosystem, including public and private financial institutions. Detailed mapping of financial actors helped identify how climate adaptation is currently integrated, what barriers exist, and where opportunities for investment lie. Financial actors included businesses from micro, small, and medium-sized enterprises (MSMEs) to large corporations, regulators, development banks, multilateral climate funds, domestic government programs, credit unions, insurers, bilateral donors, civil society, philanthropy, central banks, and so on. As an example of such evaluation, discussions with local financial institutions in Bhutan helped shape the financial matchmaking step and identified needs for technical assistance and credit mechanisms.

Recognizing the diversity of the private sector was essential for developing a comprehensive picture. For example, in Indonesia, several national commercial banks play a crucial role in facilitating access to credit for MSMEs involved in the agriculture sector. The analysis indicated that preparing investment in the national climate-smart agriculture financing scheme for these enterprises would require the involvement of a large number of agencies and financial institutions.

Finally, the process highlighted the importance of engaging development partners from the outset. For example, in Cambodia, strong early collaboration with ADB sector teams enabled the rapid integration of adaptation investment packages into the ADB project pipeline. The process also revealed a broader need for improved coordination among development partners on adaptation, particularly as countries transition from planning to implementation.

Country-led approaches yield coherence in adaptation investment planning across different actors, sectors, geographies, and governance levels.

Once key actors were identified, governments leveraged or established institutional arrangements that facilitated regular face-to-face engagement throughout the preparation of investment-ready adaptation priorities. This helped build trust and ownership, maintained momentum, and ensured transparency through ongoing communication, feedback, and collaboration at critical decision points.

Institutional arrangements (Table 4) were crucial for fostering a government-led, coordinated approach to mobilizing investments, ensuring that investment priorities were harmonized across different actors, sectors, geographies, and levels of governance. This coordination not only aligned efforts but also reduced the risk of duplication and inefficiencies.

Countries were able to leverage existing institutional mechanisms to facilitate this collaboration.

In some contexts, new mechanisms were created to fill coordination gaps. In Bhutan, for example, the establishment of an interministerial working group brought together ministries and agencies responsible for various aspects of water management—including agriculture, water and sanitation, energy, and flood management—thereby strengthening collaboration and coherence in adaptation investment planning. It fostered strong government ownership of the overall process, resulting in a unified, nationally owned plan.

Like Bhutan, Cambodia established a Technical Advisory Group composed of diverse line ministries to co-develop the climate adaptation investment plan. Improved coordination among government agencies and strong integration with existing plans helped align government and investor interests.

Table 4: Institutional Arrangements Used in the Six Countries to Coordinate the Investment Planning Process

Country	Thematic focus	Lead ministry	Institutional arrangements
Armenia	National horticulture development	Ministry of Economy with involvement of Ministry of Finance and Ministry of Environment	Leveraged the Interagency Council on Climate Change, headed by the Deputy Prime Minister
Bhutan	National water sector development	Initiated by Ministry of Finance and the technical work led by an Inter-Agency Working Group led and coordinated by the Department of Water, Ministry of Energy and Natural Resources	Established a new cross government working group chaired by Ministry of Energy and Natural Resources' Department of Water, who holds overall responsibility for national water management
Cambodia	Agriculture development for a special economic zone (comprising four provinces in the northwest)	Ministry of Economy and Finance	Established a new Technical Advisory Group composed of relevant line ministries and led by the Ministry of Economy and Finance
Indonesia	Financing scheme for climate-smart agriculture, for micro, small and medium-sized enterprises	Ministry of Finance with involvement of Ministry of Agriculture, Financial Services Authority, and Ministry for Economic Affairs	Cross-sector collaboration led by the Ministry of Finance with various line ministries and agencies
Mongolia	Water retention ponds at the national level	State Water Agency, Ministry of Environment and Climate Change with involvement from the Ministry of Food, Agriculture and Light Industry, the Ministry of Finance, the Ministry of Economy and Development, the National Hydrometeorological Service	Cross-sector collaboration led by State Water Agency, Ministry of Environment and Climate Change. Use of public investment programming systems to coordinate investment planning
Nepal	Agriculture development in Madhesh province	Ministry of Agriculture and Livestock Development with involvement from the provincial Ministry of Land Management, Agriculture and Cooperative	Strong institutional collaboration across governance levels with federal leadership, coordinating with provincial ministries, and technical inputs from the Nepal Agricultural Research Council and its subcenters and by partners such as Food and Agriculture Organization of the United Nations ensured alignment with strategic priorities and strengthened institutional ownership and aiming to feed into planning commission's project banks at national and provincial levels

Source: Authors.

In Mongolia and Nepal, public investment and financial management systems were used to coordinate adaptation investment planning. For instance, in Nepal, strong institutional collaboration across governance levels—including federal and provincial ministries and technical agencies such as the Nepal Agricultural Research Council—enhanced the design of the adaptation investment packages. This coordination also created opportunities to integrate these investments into the national and provincial project banks and the medium-term expenditure framework.

Engagement with financial institutions further helped identify ways to align investments with the national green taxonomy and explore innovative financial instruments and mechanisms, such as green bonds, ADB's proposed Green and Resilient Financing Facility in Nepal, and the Business Investment Financing Track under the Global Agriculture and Food Security Program. The latter aims to mobilize affordable climate finance for agri-food MSMEs through multilateral development banks.

In Cambodia, opportunities for coordination and integration were further improved through the delivery of capacity strengthening workshops. These included workshops each targeting a relevant sector ministry (agriculture and water), and the finance ministry, as well as a cross-government capacity strengthening workshop on adaptation financing. These workshops were also used to help prioritize the key investment packages and sources of finance.

Adaptation investment planning requires specialized skills and adequate human and financial resources.

In each country, the preparation of adaptation priorities for investment relied on specialized skills and adequate human and financial resources. For example, and as noted earlier, it required resources to ensure deeper, broader, and sustained engagement with key actors, strong moderation, and communication skills. At the same time, it was also important to have the capacity to conduct specific analysis such as more focused climate diagnostics linked to specific adaptation priorities or investment packages, more detailed cost–benefit analyses, and targeted finance matchmaking. This means that the CAIP process, as a new approach, requires ongoing awareness-raising and capacity and institutional strengthening across line ministries and departments, as well as technical assistance.

Experience from applying the CAIP framework shows that preparing the adaptation priorities for investment readiness in a sector or region requires significant resources over an extended time frame. This depends on many factors such as the quality of climate assessments available, the need for deeper pre-feasibility assessments, and the status of the adaptation planning, including the level of awareness and coordination on adaptation across government.

In each country, the process was supported by senior international experts specializing in climate risk assessment, sector adaptation, climate economics, and climate finance. These experts worked across all six countries, which facilitated valuable cross-pollination of ideas and good practices. National experts engaged in each country were key to co-leading the process and ensuring a tailored approach to country needs and priorities. Across the countries, substantial time was spent to sensitize ministries to the approach and to gain buy-in and ownership so that the approach could be sustained and institutionalized. In Nepal, active engagement across all levels of government—along with the participation of the private sector, financial institutions, research organizations, civil society, and local actors—facilitated cross-learning and coordination, leading to the effective design and strong ownership of the CAIP. This process was supported through consultations in Kathmandu and Madhesh Province, key informant interviews, and a series of validation workshops. Robust data collection also played a central role, including the review of existing policies and plans, field surveys for key value chains, and the compilation of relevant secondary data. Together, these efforts strengthened the overall foundation of the investment planning process.

The process further underscored the need across all countries for stronger coordination and harmonization among development partners on adaptation investment planning to ensure that efforts would provide developing countries with coherent technical and financial support. Implementing adaptation investment plans will require ongoing coordination and support, potentially through project preparation facilities and country climate platforms, with strong government ownership and coordinated backing from development partners.

5 CONCLUSIONS

There is growing consensus about the activities and types of information necessary to prepare investment-ready adaptation priorities. These steps and requirements have often been overlooked, resulting in a fragmented landscape of small, isolated, and duplicative adaptation projects, which in turn calls into question the added value of the NAP process. Clarifying what is needed to develop investment-ready adaptation priorities is essential for improving the effectiveness of NAP processes, addressing the adaptation finance gap, and tracking progress under the global goal on adaptation.

This analysis reviewed the experiences of six countries preparing investment-ready climate adaptation priorities and identified nine key lessons. Successfully mobilizing adaptation investments requires balancing ambition with pragmatism and building granularity through iteration. Strengthening the climate evidence base and rationale helps prioritize investments and identify financing sources. Comprehensive adaptation packages should address multiple climate risks at the system level with diverse investment types. Evaluations of the benefits of investing in adaptation can use qualitative assessments and proxies, while simple heuristics can aid finance matchmaking. It is also vital to broaden and deepen engagement with key actors, foster a country-led coordinated approach, and ensure adequate financing for the journey to investment readiness.

As countries accumulate experience and progress through successive NAP cycles, these activities and information could be progressively integrated into updated NAPs or other strategic documents.

Governments and other actors are still in the early stages of understanding how to achieve adaptation investment readiness effectively. Therefore, it is crucial to document and assess the impacts of technical assistance provided through the CAIP and similar initiatives. A key question is whether and how these CAIPs will help countries secure diverse sources of finance. Capturing results and lessons from these experiences is essential for enabling countries to mobilize investments for adaptation more effectively.

As of this report's publication, governments across the six countries have either endorsed or are in the process of endorsing the climate adaptation investment plans, highlighting government ownership. In several countries the plans are being used to promote a more harmonized approach within governments and among development partners, such as in Bhutan.

The CAIP process has also led to the mobilization of finance in countries. For example, this includes new adaptation grant finance in Cambodia through the ADB Asian Development Fund. It has also led to the inclusion of some adaptation investment packages in ADB country pipelines, as well as opportunities with other sources of finance.

The immediate next step for the six countries in focus is to maintain the momentum and secure financing for their CAIP by undertaking detailed feasibility studies of investment packages and developing project proposals or concept notes. Looking ahead, more analysis is needed on who would benefit from the identified investment packages.

Climate-resilient horticulture in Armenia includes investment in adaptation activities like water-ponds to collect rainwater and restore groundwater; use of drip irrigation and hail nets to address climate impacts to horticulture. © Nanki Kaur

Future work should explore how integration of gender equality, disability inclusion, and other social considerations can be strengthened in the process of preparing investment-ready climate adaptation priorities. This is crucial to avoid harm and ensure that investments in adaptation address the needs of people in situations of high vulnerability to climate risks. As such, engagement must extend beyond ministries of finance, sector agencies, and financial actors to ensure civil society organizations are involved at all levels—not only in identifying adaptation priorities but also in refining and justifying investments, particularly when these are expected to generate low financial returns. This will help to address the risk of overlooking vulnerable populations during the transition from planning to implementation and the need to pursue inclusive adaptation actions.

Further work is also needed to better understand how countries can more effectively leverage domestic resources, including through fiscal tools such as taxes, public procurement, and subsidies, to secure domestic finance for adaptation.
In addition, ongoing exploration is essential to identify ways countries can attract increased private sector investment through blended finance and policy reforms.

Forest watershed management is critical to ensure a sustainable water supply. A team from Bhutan's Department of Water monitors the regeneration of an area previously scorched by forest fire. © Sonam Chophel

REFERENCES

ADB. 2017. Guidelines for the Economic Analysis of Projects. https://www.adb.org/documents/guidelines-economic-analysis-projects

ADB. 2024a. *Climate-Resilient Fiscal Management: Experience from Southeast Asia*. https://www.adb.org/publications/climate-resilient-fiscal-management-southeast-asia.

ADB. 2024b. *Climate Adaptation Investment Planning. A Program to Bridge the Gap between Climate Adaptation Planning and Financing*. https://www.adb.org/sites/default/files/publication/927306/caip-brochure.pdf.

ADB. 2025a. Armenia: Improved Decision-Making for Climate Resilient Development in Asia and the Pacific: Climate Adaptation Investment Planning-Armenia Sectoral Report. Consultants' report. (TA 6629-REG). https://www.adb.org/sites/default/files/project-documents/54412/54412-001-tacr-en.pdf.

ADB. 2025b. Indonesia: Improved Decision-Making for Climate Resilient Development in Asia and the Pacific: Climate Adaptation Investment Plan for Green KUR for Climate-Smart Agriculture. Consultants' report. (TA 6629-REG). https://www.adb.org/sites/default/files/project-documents/54412/54412-001-tacr-en_0.pdf.

ADB. Forthcoming-a. Economic Analysis in the Climate Adaptation Investment Planning. CAIP Working Paper Series.

ADB. Forthcoming-b. Financial Matchmaking in the Climate Adaptation Investment Planning Process. CAIP Working Paper Series.

ADB. Forthcoming-c. Undertaking Climate Diagnostics for Climate Adaptation Investment Planning. CAIP Working Paper Series.

ADB. Forthcoming-d. Bhutan: Climate Adaptation Investment Plan for the Water Sector. Consultants' report. (TA 6933-REG).

ADB. Forthcoming-e. Cambodia: Climate Adaptation Investment Plan for the Agriculture Sector in Economic Pole 4. Consultants' report. (TA 6933-REG).

ADB. Forthcoming-f. Mongolia: Improved Decision-Making for Climate-Resilient Development in Asia and the Pacific: Climate Adaptation Investment Plan for Climate-Smart Water Ponds in Mongolia. Consultants' report. (TA 6629-REG).

ADB. Forthcoming-g. Nepal: Climate Adaptation Investment Plan for the Agriculture Sector in Madhesh Province. Consultants' report. (TA 6933-REG).

ADB and the International Institute for Sustainable Development. 2025. *Climate Change Adaptation Investment Planning Newsletter*. Issue No. 1, July. https://napglobalnetwork.org/wp-content/uploads/2025/07/Newsletter_caip01_COMPRESSED-2025-07-03-Final.pdf.

Brandon, C., B. Kratzer, A. Aggarwal, and H. Heubaum. 2025. *Strengthening the Investment Case for Climate Adaptation: A Triple Dividend Approach*. World Resources Institute (WRI). https://www.wri.org/research/climate-adaptation-investment-case.

Climate Policy Initiative (CPI). 2025. *Global Landscape of Climate Finance 2025.* https://www.climatepolicyinitiative.org/publication/global-landscape-ofclimate-finance-2025.

Coalition of Finance Ministers for Climate Action. 2023. *Strengthening the Role of Ministries of Finance in Driving Climate Action. A Framework and Guide for Ministers and Ministries of Finance. Final Report.* World Resources Institute. https://www.wri.org/research/climate-adaptation-investment-case.

Dekens, J. 2023. *Using Climate Risk Assessment to Measure Adaptation Success at the National Level. Preliminary Lessons from 12 Countries.* NAP Global Network, International Institute for Sustainable Development. https://napglobalnetwork.org/resource/climate-risk-assessment-measure-adaptation-success/.

Hammill, A., Dazé, A., and Dekens, J. 2020. *The National Adaptation Plan (NAP) Process: Frequently Asked Questions.* NAP Global Network, International Institute for Sustainable Development. https://napglobalnetwork.org/2019/12/the-national-adaptation-plan-nap-process-frequently-asked-questions/.

Hernández, M., C. Ledwell, and G. J. Yang. 2025. *Finance for National Adaptation Plan Processes. What Can We Learn from Countries' National Adaptation Plans?* NAP Global Network, International Institute for Sustainable Development. https://napglobalnetwork.org/resource/finance-for-nap-processes/.

Intergovernmental Panel on Climate Change (IPCC). 2023. Summary for Policymakers. In *Climate Change 2023: Synthesis Report.* Contribution of Working Groups I, II and III to the Sixth Assessment Report of the Intergovernmental Panel on Climate Change [Core Writing Team, H. Lee and J. Romero (eds.)]. pp. 1–34.

NDC Partnership and Green Climate Fund. 2024. *Climate Investment Planning and Mobilization Framework.* https://ndcpartnership.org/sites/default/files/2024-12/climate-investment-planning-and-mobilization-framework.pdf.

Organisation for Economic Co-operation and Development (OECD). 2024. *Climate Adaptation Investment Framework.* Green Finance and Investment, OECD. https://www.oecd.org/en/publications/climate-adaptation-investment-framework_8686fc27-en.html.

Planetary Boundaries Science (PBScience). 2025. Planetary Health Check 2025. Potsdam Institute for Climate Impact Research (PIK).

Pokhrel, A. 2024. Decentralization Dilemmas: Lessons from Nepal's Federal Transition. *The Annapurna Express*, 5 May. https://theannapurnaexpress.com/story/48703/.

United Nations. 2025. *Sevilla Commitment.* Fourth International Conference on Financing for Development. 30 June–3 July. https://unctad.org/system/files/official-document/aconf227-2025-l1_en.pdf.

United Nations Environment Programme (UNEP). 2023. *Adaptation Gap Report 2023: Underfinanced. Underprepared.* https://www.unep.org/resources/adaptation-gap-report-2023.

UNEP. 2024. *Adaptation Gap Report 2024: Come Hell and High Water.* https://www.unep.org/resources/adaptation-gap-report-2024.

UNEP. 2025. *Adaptation Gap Report 2025: Running on Empty.*

United Nations Framework Convention on Climate Change (UNFCCC). 2011a. Report of the Conference of the Parties on its sixteenth session, held in Cancun from 29 November to 10 December 2010. Addendum. Part two: Action taken by the Conference of the Parties at its sixteenth session (FCCC/CP/2010/7/Add.1). https://unfccc.int/sites/default/files/resource/docs/2010/cop16/eng/07a01.pdf.

UNFCCC. 2011b. National Adaptation Plans. Decision 5/CP.17. https://unfccc.int/files/adaptation/cancun_adaptation_framework/national_adaptation_plans/application/pdf/decision_5_cp_17.pdf.

UNFCCC. 2025a. *Report on the Baku to Belém Roadmap to 1.3T*. https://unfccc.int/sites/default/files/resource/Relatorio_Roadmap_COP29_COP30_EN_final.pdf.

UNFCCC. 2025b. *The NAP Technical Guidelines Updated Technical Guidelines for the Process to Formulate and Implement National Adaptation Plans*. LDC Expert Group. 250812 UNFCCC LEG UPDATED TECHNICAL GUIDELINES.pdf.

UNFCCC. 2025c. Matters Relating to the Global Goal on Adaptation (FCCC/PA/CMA/2025/L.25). https://unfccc.int/sites/default/files/resource/cma2025_L25_adv.pdf.

Watkiss, P., and R. A. Betts. 2021. Method. In *The Third UK Climate Change Risk Assessment Technical Report* [R. A. Betts, A. B. Haward, and K. V. Pearson (eds.)]. Prepared for the Climate Change Committee, London. https://www.ukclimaterisk.org/publications/technical-report-ccra3-ia/chapter-2/.

World Bank, Government of Nepal, and Food and Agriculture Organization of the United Nations. 2021. *Climate Smart Agriculture Investment Plan: Nepal*. Climate-Smart Agriculture Investment Plan: A Study Implemented in Collaboration with Government of Nepal, Food and Agriculture Organization and World Bank.

www.ingramcontent.com/pod-product-compliance
Lightning Source LLC
Chambersburg PA
CBHW041800220926
51847CB00014B/268
9789292776992